YOUR KNOWLEDGE HAS VALUE

- We will publish your bachelor's and
  master's thesis, essays and papers

- Your own eBook and book -
  sold worldwide in all relevant shops

- Earn money with each sale

Upload your text at www.GRIN.com
and publish for free

Bibliographic information published by the German National Library:

The German National Library lists this publication in the National Bibliography; detailed bibliographic data are available on the Internet at http://dnb.dnb.de .

This book is copyright material and must not be copied, reproduced, transferred, distributed, leased, licensed or publicly performed or used in any way except as specifically permitted in writing by the publishers, as allowed under the terms and conditions under which it was purchased or as strictly permitted by applicable copyright law. Any unauthorized distribution or use of this text may be a direct infringement of the author s and publisher s rights and those responsible may be liable in law accordingly.

Imprint:

Copyright © 2015 GRIN Verlag, Open Publishing GmbH
Print and binding: Books on Demand GmbH, Norderstedt Germany
ISBN: 978-3-668-17327-9

This book at GRIN:

http://www.grin.com/en/e-book/317273/the-german-nuts-2-region-of-arnsberg-an-economic-analysis-of-eurostat

Lennart Voss

The German NUTS-2 region of Arnsberg. An economic analysis of EUROSTAT regional data

GRIN Publishing

GRIN - Your knowledge has value

Since its foundation in 1998, GRIN has specialized in publishing academic texts by students, college teachers and other academics as e-book and printed book. The website www.grin.com is an ideal platform for presenting term papers, final papers, scientific essays, dissertations and specialist books.

Visit us on the internet:

http://www.grin.com/

http://www.facebook.com/grincom

http://www.twitter.com/grin_com

List of Content

Introduction

This report will be about the German NUTS-2 region of Arnsberg. It will include a brief introduction to the region, stating the most important facts and information. This acquired basic knowledge will then be transferred to the more specific part of the assignment, the analysis of regional data. In this case, the used tool is called EUROSTAT. It is the European Union's official source for statistics and data concerning the different member states. The Main part therefore includes charts, diagrams and tables regarding different topics as population, wealth, employment or education. Every chart will be briefly described and analyzed. Furthermore, data of the German state and the European Union is used in addition to the regional NUTS-2 data to compare on national and international level. In continuation to this, location theories and the importance of economic key drivers and agents are discussed and applied to the NUTS-2 region of Arnsberg. The assignment is then rounded-off by a conclusion and the list of used sources.

The Region of Arnsberg

Facts

The region of Arnsberg and its eponymous city is located within the western German federal state of North Rhine-Westphalia (NRW). It is one of five administrative districts within NRW, containing counties and cities like Dortmund, Bochum, Hagen, Unna and the Hochsauerland district. Looking at the map of NUTS-2 regions, it is defined as "DEA5". Arnsberg has a population on about 3.5 million

people in an area of approximately 8002km², measured in the end of 2013[1].

Picture 1.
Illustration of the German NUTS2 Regions

---

[1] https://en.wikipedia.org/wiki/Arnsberg_(region)

Picture 1: http://ec.europa.eu/eurostat/documents/3859598/5916917/KS-RA-11-011-EN.PDF/2b08d38c-7cf0-4157-bbd1-   1909bebab6a6?version=1.0

Infrastructure

The regions infrastructure is really good, as it is in most parts of Germany. It covers the Airport of Dortmund and the airport of Paderborn. The highways A1 and A2 are reachable in about 15 minutes via the A46 and the A445 going from the city of Arnsberg.

Moreover, the city of Arnsberg is reachable via one of its 3 railway stations or by long-distance busses. As an industrial institution, it is no problem to acquire needed goods in the Region of Arnsberg. If you are comparing the regions accessibility of highways on a national level, it is way above the German average, which is pretty high already. The interactive atlas from the German Newspaper "Sueddeutsche" states, that the region of Arnsberg covers 55.33km highway pr. 1000km^2, compared to the German average of 33.6km pr. 1000km^2.[2]

Picture 2. Infrastructure of Arnsberg

Industry

Everything is very well connected and highly industrialized. In addition to that, Arnsberg is situated around mineral resources in the form of granite, natural gas and coal. Those are being mined in - and around the cities.

The local economy catalogue from the city of Arnsberg states, that the manufacturing of luminary, paper and wood, just as safety equipment within cars are all playing vital roles regarding the regions healthy industrial sector. Moreover, the river "Ruhr" is floating through the region of Arnsberg, and it has been providing water for the people and the industry throughout the centuries.[3]

[2] http://www.sueddeutsche.de/politik/interaktiver-atlas-so-lebt-europa-1.1615912
[3] http://www.arnsberg.de/informationen/verkehrsverbindungen.php,
http://www.wfa-arnsberg.de/cms/upload/downloads/wirtschaftsbroschuere.pdf, page 22-23
Picture 2: http://www.wfa-arnsberg.de/cms/upload/downloads/wirtschaftsbroschuere.pdf, page 6

Analysis of Regional Data

A. Make a diagram to compare the level and development of the region's GDP (in PPS) with
    that of the whole country.

This diagram based on the data
of EUROSTAT is showing the
history and development of the
Gross Domestic Product within
the region of Arnsberg, in
comparison to the national GDP,
indicated by the purchasing
power standard per inhabitant.
Sadly, the region of Arnsberg
only had data on GDP in PPS for
the period of 2010-2012 on
EUROSTAT. Even if a broader

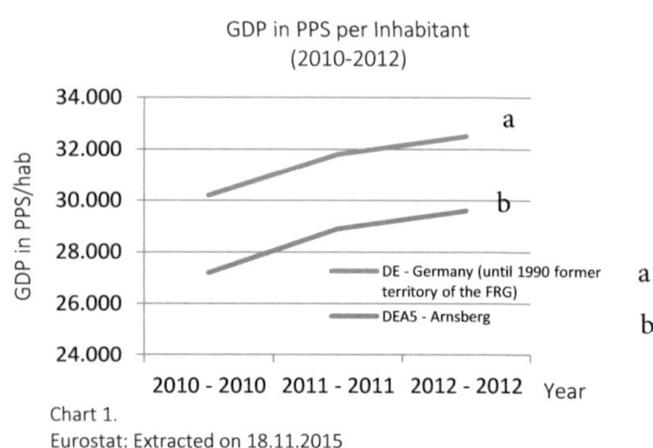

Chart 1.
Eurostat: Extracted on 18.11.2015

lapse of time would be of interest, the available data still provides serviceable tools. The
pathway of both indicators is very similar; still, the numbers of the German state are
considerably higher than Arnsberg's throughout the time taken. 29.600 PPS/hab compared to
32.500 PPS/hab in 2012. This is indicating that the region of Arnsberg is less effective in
comparison to the average Germany.

B. Make a diagram to compare the population development of the region with that of the
    whole country (use indexes).

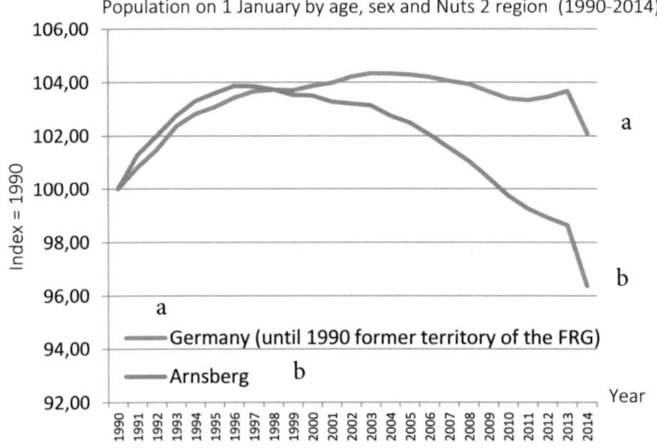

Chart 2.
EUROSTAT: Extracted on 18.11.2015

This Chart is showing the significant change in population within Arnsberg and the rest of Germany. The illustration requires the use of indexes, as the original numbers are differing too much to show them in one graph. Germany is having about 82.5 million inhabitants, while Arnsberg is on about 3.5 million. [4] Starting in 1990, both graphs are increasing equally until the year of 1997, where the graph of Arnsberg starts decreasing for the first time, while the graph for the whole of Germany stabilises around 1998-2000, just to slowly increase again up to the year of 2005, where it starts to decrease again on a minor scale. So after all, the German graph is increasing from 100% to 104%, and stays around that number up until the year of 2013, were it decreases kind of rapidly from 103,7% to 102%. This might be caused by the financial crisis in 2007. It has been the trend in Germany for a few years; the population is decreasing due to the high costs connected to children and the increasing percentage of women working. Germany is facing the problems of a very old population. German inhabitant's rate of birth is too low; to make sure that a decent pension for retired generations is secured. [5] The graph for Arnsberg is decreasing even further, after stabilizing around 103% 1996-2003. It has taken a decrease from 103% to 98.5% from 2003-2013, and then falling even more after 2013 (just as the German average), reaching its lowest point around 96% in 2014. Starting in 1990, Arnsberg had a total decrease in population by nearly 4%, while the German average still has increased 2% overall, currently having a negative tendency. Furthermore, Arnsberg is having about 15% school dropouts, while Germany´s average is on about 12%in 2013[6]. This could really become a problem when the local industry needs to fill the gaps occurring within the manufacturing industry. So the region has to ensure that people will come to keep up the current labour.

[4] http://www.worldometers.info/world-population/germany-population/
[5] https://www.cia.gov/library/publications/the-world-factbook/geos/gm.html
[6] http://www.sueddeutsche.de/politik/interaktiver-atlas-so-lebt-europa-1.1615912

C. Make a triangle diagram to compare the education level in the region, in the country and
    in EU-28.

The triangle diagram gives us some insight on the educational level of Arnsberg in comparison to the 28 EU States and the German state. The chart is including people of 15 years and older and shows as following:

Level 0-2: Lower Secondary education or less.

Level 3-4: Upper Secondary, non-tertiary education.

Level 5-8: Tertiary Education.

While Germany has got 13% of the people with lower secondary education, it has a 60% within the upper secondary - and 27% through tertiary education.

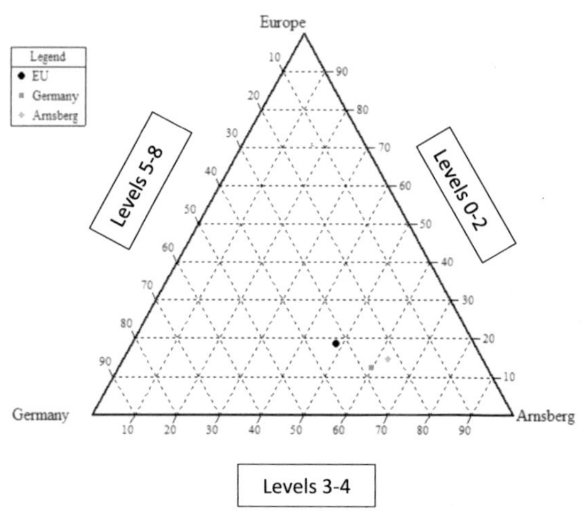

Chart 3.
EUROSTAT: Extracted on 14.12.2015

Meanwhile, Arnsberg does have 15% with a lower secondary education, 63% with upper secondary and 22% with a tertiary educational level. Europe appears to have 19% within lower secondary, 49% on upper secondary and 32% people with tertiary education.

It shows us that Arnsberg does have more people than average Germany, when it comes to a low education; still, Europe is having the highest percentage of lower educated citizen. For the time being, Arnsberg does have the top percentage when it comes to upper secondary education. One of the reasons could be that this is enough to acquire labour within the big industry sector in Arnsberg. Europe's percentage in the upper secondary section is significantly lower. They do have the highest amount of people with a tertiary education though, whilst also having the highest amount of people within lower secondary education or lower. The education level in Arnsberg is pretty good; still, the region should tackle their problems regarding population decrease and level of education, if they want to keep the progress of industry, manufacturing and innovation going.

D. Make a diagram (stacked columns) to compare the employment in the three main
    business sectors in the region with that in the whole country.

This diagram shows the transition
of German society towards the
service sector. The primary sector
is so highly modernized, that the
percentage of people enforced is
close to nothing, while the sector
remains productive. In all highly
developed countries, the vast
majority of people are working
within the service sector.
Germany still has a lot of labour in
the industrial sector though,

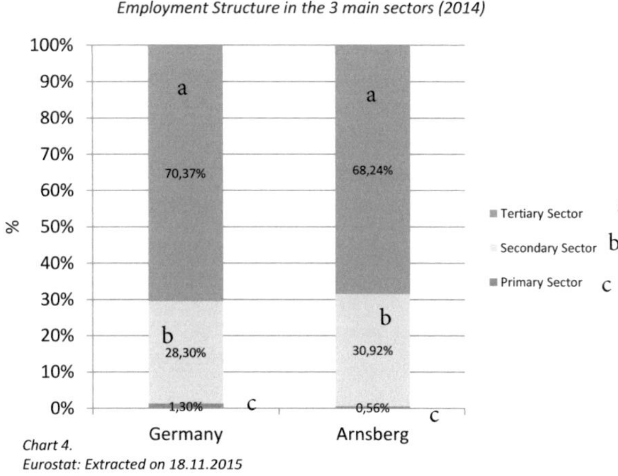

Chart 4.
Eurostat: Extracted on 18.11.2015

reasoned by their highly productive manufacturing institutions in some areas within the
country. As an example, France is only employing about 19.4% of their workforce within the
secondary sector[7].

The "Ruhrgebiet", including Arnsberg is counted as one of those areas focussing on
manufacturing and industry This explains the above average percentage of 30.9%.

---

[7] http://de.statista.com/statistik/daten/studie/167247/umfrage/anteile-der-wirtschaftssektoren-am-
bruttoinlandsprodukt-in-frankreich/

E. Make a regression analysis between the region´s employment and its GVA in business sectors (10 sectors).

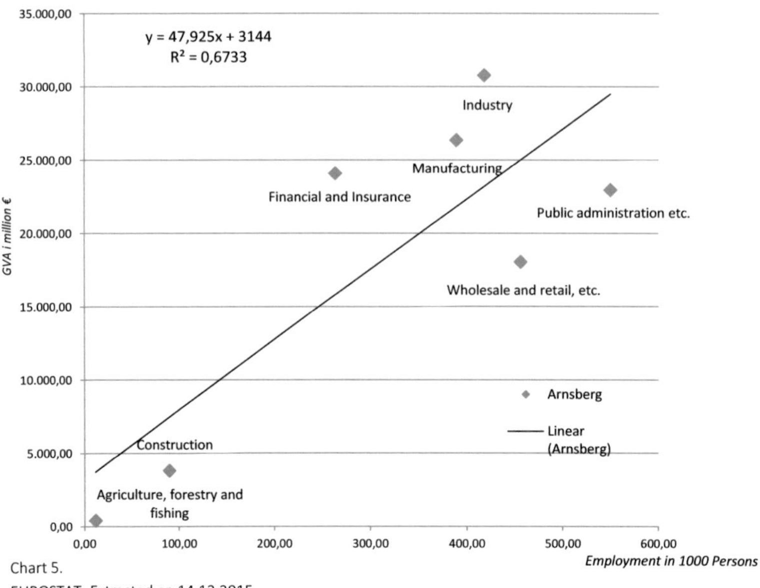

Chart 5.
EUROSTAT: Extracted on 14.12.2015

Chart 5 shows us both the GVA and the employment in business sectors in relation to each other. The formula "y=47,925x+3144 is illustrating this by generating a trendline between some of the regions sub sectors. For example, if we take a look at Industry, Manufacturing and Financial Insurance, they do generate quite an amount of the GVA considering that they do have less employees than Public Administration and Wholesale and Retail etc. The generated R^2 meanwhile shows a value of 0.6733, while a value of 0 indicates no linear relation, and a value of 1 indicates a perfect linear relation. [8]

---

[8] https://www.inwt-statistics.de/blog-artikel-lesen/Bestimmtheitsmass_R2-Teil5.html

F. Make a table with the region´s top-ten business sectors based on location quotients. Data as well as calculations should be included in the table.

Employment in 2012		Employed Persons		Share in %		
No.	Top 10 Sub Sectors in DE45 Arnsberg	DE45 Arnsberg	DE Germany	DE45	DE	LQ DE45
1.	Manufacturing	353.076	1.066.733	15,28	1,84	8,30
2.	Administrative and support service activities	106.715	815.805	4,62	1,41	3,28
3.	Retail trade, except of motor vehicles and motorcycles	146.163	1.153.761	6,33	1,99	3,18
4.	Accommodation and food service activities	58.807	486.136	2,55	0,84	3,04
5.	Professional, scientific and technical activities	62.453	599.903	2,70	1,03	2,61
6.	Manufacture of fabricated metal products, except machinery and equipment	83.043	812.893	3,59	1,40	2,56
7.	Construction	62.543	614.864	2,71	1,06	2,55
8.	Specialised construction activities	48.133	477.245	2,08	0,82	2,53
9.	Manufacture of basic metals	52.149	531.921	2,26	0,92	2,46
10.	Food and beverage service activities	46.272	477.245	2,00	0,82	2,43
...	...	...	...	...	...	...
	Total	2.310.685	57.973.559	100	100	1

Chart 6.
Eurostat: Extracted on 14.12.2015

The sixth chart is presenting the popularity of some specific sub-sectors within the region of Arnsberg, in connection to the popularity of the sector on national scale. The "Location Quotient" is calculated by dividing the share of employed persons in the Region with the share of the German state in one explicit sector. The LQ then tells us something about the popularity of the sector. The manufacturing sub-sector has a LQ on 8.30, so this tells us that the manufacturing sector is 8.3 times more popular in Arnsberg, compared to the rest of Germany. Manufacturing is far beyond the other sectors looking at the LQ. About every third person employed in the manufacturing sector is employed in the Region of Arnsberg in 2012! Overall it's pretty clear that manufacturing of different things is of high interest. It really fits the proportional distribution of the employment structure from Chart 4.

Importance of Key Agents

A. Firm

A lot of big industrial giants and businesses are located around the so-called "Ruhrpott", most of them in the bigger cities like Essen, Düsseldorf or Cologne. Still, there are many important firms within the region of Arnsberg, firms like "RAG Deutsche Steinkohle AG" in Herne, or the"

AOK", "ThyssenKrupp" and "Tedi" in the city of Dortmund. They are securing work of thousands of people within the Region.[9]

B. Labour

North Rhine-Westphalia is known for its high rates of unemployment, Dortmund has had an unemployment rate of 13.2% in the year of 2013[10]. The manufacturing and industrial sector in Arnsberg is suffering from the same problems as in many other regions. Globalization leads to outsourcing of processes due to cheaper wages. Furthermore, temporary workers from the east are way cheaper, which leads to unemployment. As shown in chart 3, Arnsberg is lacking when it comes to education, when comparing it to the national level.

C. State

Germany consists of 16 federal states, including federal governments involving politicians elected by the public. The state has a free and social market economy.

The book "Key concepts in Economic Geography" states, that monopolistic market situations are very dangerous, luckily, monopoles as-well as corruption are cut to a minimum in Germany.[11][12] The state is following the model of a social welfare system, trying to create equal possibilities for everyone. Unemployed citizen are receiving a certain amount of money to live, managers have to pay a minimum wages and the government is interested in helping upcoming entrepreneurs and start-ups via investments, which are serving both region and individuals at the same rate.[13]

Key Drivers of Arnsberg

A. Accessibility

Choosing from the 3 different key drivers, I am going with accessibility above Innovation. In my opinion, it appears to be the main reason for the regions success. The Rhein-Ruhr area is

---

[9] http://revier-manager.de/arbeitgeber/ruhrgebiet

[10] http://www.welt.de/regionales/duesseldorf/article123633384/Agentur-rechnet-mit-mehr-Arbeitslosen.html

[11] Yuko Aoyama, James T. Murphy, Susan Hanson: Key Concepts in Economic Geography, Page 38, SAGE Publications Ltd., London 2011

[12] http://www.transparency.org/cpi2014/results

[13] http://www.wfa-arnsberg.de/cms/upload/downloads/wirtschaftsbroschuere.pdf,Page 35, Norbert Runde

one of the hottest on the location heat map, being ranked 4th on the 2013 list of most popular logistical locations. Therefore, many smaller industrial and manufacturing businesses, which are depending on good infrastructure, labor and logistics, are gathering in this area.[14] As stated earlier, the magnificent infrastructure of Arnsberg in the form of airports, highways, railway-stations and logistic centers is above the already high average of the German state.[15] Furthermore, the river Ruhr is giving access to drinking - and industrial water. Looking at the IT accessibilities, it appears that Arnsberg is very well connected all around the region. The Governor of Arnsberg is actually talking about creating free accessible Wireless Lan for the Inhabitants of the city.[1617]

B. Innovation

Arnsberg is depending on continuing innovation, as the region is mostly about industry. The local companies are investing to constantly modernize their ideas and processes to live up to its international reputation and the global competition. Without technological innovation, the big industrial sector wouldn't be able to keep up with the global market prices and the public demand. So without a doubt, innovation does play a giant role in Arnsberg, still I would rate accessibility more important.[18]

C. Entrepreneurship

The Global Entrepreneurship Monitor (GEM) states, that a good physical infrastructure is one of the key assets when setting up a new business. Still, Germany is not entirely maximizing the use of its entrepreneurial potential, which they got due to magnificent infrastructure and governmental support when it comes to setting up a new business. In 2014, 40% of the population stated (excluding individuals involved in entrepreneurial actions), that they are not setting up a business due to their fear of failure in the progress of doing so. GEM concludes that the Germans tendency of risk aversion has a decelerating effect when it comes to the countries entrepreneurial growth level. The Total early stage Entrepreneurial Activity (Individuals between 18 and 64) is at about 5.3%. There is no comparable industrialized and

---

[14] https://e-learn.sdu.dk/bbcswebdav/pid-4204385-dt-content-rid-5954954_2/courses/9858301-SB-E15/WinningLocationsEurope.pdf
[15] The region of Arnsberg, Page 2
[16] http://www.derwesten.de/staedte/arnsberg/ziel-freies-internet-fuer-arnsberger-id8717525.html
[17] http://www.britannica.com/place/Ruhr
[18] http://www.wfa-arnsberg.de/cms/upload/downloads/wirtschaftshroschuere.pdf , Page 16-21

innovative country with a TEA being significantly below the German value of 2014. Furthermore, the entrepreneurial activity among young people under the age of 25 has been decreasing enduringly during the preceding years. Another reason for the low TEA might be found observing the job opportunities offered to the German citizens. The chances to get a decent job are fairly good, and accordingly, the temptation to risk building up something on your own is not as high as it is in countries were job opportunities are worse. Therefore, I do rate entrepreneurship less important in my region Arnsberg.[19]

Location Theories

Location theories are attempts to explain the real world in a simplified way, by using theoretical concepts and models. [20]

Alfred Weber´s "Least Cost Theory"
Weber is writing about the different factors affecting production costs: Raw Material, transportation and labor. He is aiming at finding the cheapest place to produce goods. I think his theory is very applicable to the region of Arnsberg. The manufacturing/industrial institutions have been depending on local factors as the "Ruhr", natural resources as granite, coal and proximity of time and place. Weber mentions the purpose of agglomeration, and I think that the region of Arnsberg really is making use of its great connectivity and closeness to different hotspots. Furthermore, the need for cheap labor is embodied by the high unemployment rate, due to cheap temporary workers from Eastern Europe and outsourcing.[21]

---

[19] http://www.gemconsortium.org/country-profile/64, GEM Germany 2014 Report, Page 9, Data taken on 12.11.15 ,
http://www.gemconsortium.org/country-profile/64, Data from 2014, Taken on 12.11.15
[20] https://e-learn.sdu.dk/bbcswebdav/pid-4238568-dt-content-rid-6061460_2/courses/9858301-SB-E15/Lecture%2012%20Handouts.pdf, slide 3
[21] Yuko Aoyama, James T. Murphy, Susan Hanson: Key Concepts in Economic Geography, Page 77-80, SAGE Publications Ltd., London 2011

Socio-Cultural Issues

A. Culture

The region of Arnsberg seems to be very open minded, in October 2015, 214.000 refugees found shelter within the federal state of NRW. The official homepage of Arnsberg is promoting all kinds of refugee friendly initiatives.[22] With that being said, it is safe to say that the "Ruhrpott" is a region of hard workers, using rather few words. The long tradition of industry and coalmining is embedded in the mentality of a lot of people within the region. People are used working side by side with other religions, since the first big migration waves from Turkey and eastern European countries and workers from a lot of foreign countries arrived in the 60ies. The level of tolerance, while being conservative at times, might be one of the reasons for Germany`s economic improvement throughout the decades.

B. Gender

Germany´s politicians are always troubled with the topic of gender equality. It has been a constant theme in most of the political discussions. Throughout the years, there has always been criticism on the pay gap between men and women. Also, the percentage of women in supervisory boards is significantly lower, on around 18% in 2014. Nonetheless, Germany is still a very tolerant and equally fair country compared to some countries within the European Union. Women are working just about as much as men, and well, even the chancellor is a woman. [23]

Conclusion

Arnsberg is a very interesting region with great strength and a lot of possibilities, not only within sector of manufacturing. Great infrastructure and local proximity ensures high productivity all over Arnsberg. But there are many problems to tackle. Dealing with the decreasing rate of birth, working down the high rate of unemployment and ensuring high level education should be the way to go. Maybe the current wave of refugees will become a good chance for German regions, suffering from fading labour forces?

---

[22] http://www.bezreg-arnsberg.nrw.de/integration_migration/fluechtlinge_in_nrw/index.php
[23] http://de.statista.com/statistik/daten/studie/180107/umfrage/frauenanteil-in-den-aufsichtsraeten-der-200-groessten-deutschen-unternehmen/

List of sources

Literature

Yuko Aoyama, James T. Murphy, Susan Hanson

Key Concepts in Economic Geography

SAGE Publications Ltd., London 2011

Web Sources

https://en.wikipedia.org/wiki/Arnsberg_(region) (Last visited: 14.12.15)

http://www.sueddeutsche.de/politik/interaktiver-atlas-so-lebt-europa-1.1615912 (Last visited: 14.12.15)

http://www.arnsberg.de/informationen/verkehrsverbindungen.ph (Last visited: 14.12.15)

http://www.wfa-arnsberg.de/cms/upload/downloads/wirtschaftsbroschuere.pdf (Last visited: 14.12.15)

http://www.worldometers.info/world-population/germany-population/ (Last visited: 14.12.15)https://www.cia.gov/library/publications/the-world-factbook/geos/gm.html (Last visited: 14.12.15)

https://www.inwt-statistics.de/blog-artikel-lesen/Bestimmtheitsmass_R2-Teil5.html (Last visited: 16.12.15)

http://revier-manager.de/arbeitgeber/ruhrgebiet (Last visited: 15.12.15)

http://www.welt.de/regionales/duesseldorf/article123633384/Agentur-rechnet-mit-mehr-Arbeitslosen.html (Last visited: 16.12.15)

http://www.transparency.org/cpi2014/results (Last visited: 16.12.15)

https://e-learn.sdu.dk/bbcswebdav/pid-4204385-dt-content-rid- (Last visited: 5954954_2/courses/9858301-SB-E15/WinningLocationsEurope.pdf (Last visited:15.12.15

http://www.derwesten.de/staedte/arnsberg/ziel-freies-internet-fuer-arnsberger-id8717525.html (Last visited: 24.11.15)

http://www.britannica.com/place/Ruhr (Last visited: 16.12.15)

http://www.bezreg-arnsberg.nrw.de/integration_migration/fluechtlinge_in_nrw/index.php (Last visited: 16.12.15)

http://www.gemconsortium.org/country-profile/64 (Last visited: 15.12.15)

https://e-learn.sdu.dk/bbcswebdav/pid-4238568-dt-content-rid-6061460_2/courses/9858301-SB-E15/Lecture%2012%20Handouts.pdf (Last visited:17.12.15)

<u>Pictures</u>

Picture 1: http://ec.europa.eu/eurostat/documents/3859598/5916917/KS-RA-11-011-EN.PDF/2b08d38c-7cf0-4157-bbd1-1909bebab6a6?version=1.0

Picture 2: http://www.wfa-arnsberg.de/cms/upload/downloads/wirtschaftsbroschuere.pdf, page 6

<u>Data</u>

Chart 1-6 extracted from EUROSTAT:
http://ec.europa.eu/eurostat/web/regions/data/database

http://de.statista.com/statistik/daten/studie/167247/umfrage/anteile-der-wirtschaftssektoren-am-bruttoinlandsprodukt-in-frankreich/ (Last visited: 15.12.1)5
http://de.statista.com/statistik/daten/studie/180107/umfrage/frauenanteil-in-den-aufsichtsraeten-der-200-groessten-deutschen-unternehmen/ (Last visited: 15.12.15)
http://www.sueddeutsche.de/politik/interaktiver-atlas-so-lebt-europa-1.1615912 (Last visited: 17.12.15)

YOUR KNOWLEDGE HAS VALUE

- We will publish your bachelor's and
  master's thesis, essays and papers

- Your own eBook and book -
  sold worldwide in all relevant shops

- Earn money with each sale

Upload your text at www.GRIN.com
and publish for free